Books by Dennis M Keating

The Olympics:
An Unauthorized Unsanctioned History
*
Charlie Whitman
Was a Friend of Mine
*
Ena Road
*
The Fulda Gap
*
A Chicago Tale
*
Black Lahu
*
Poetry for Men

Fulda Gap

A Cold War Standoff

Dennis M Keating

This book was created by
the Golden Sphere team,
in coordination with the Honolulu Guy,
Dennis M Keating

Golden Sphere
GS
www.goldensphere.com

The Author

Dennis M Keating

The Honolulu Guy

DEDICATION

To the Men and Women who Defend Our Freedom

ACKNOWLEDMENTS

Thanks to

Professor Steven Taylor Goldsberry
My Mentor

Paula Marie Fernandez
and Hikari Kimura
For Artwork and Maps

Gail M Baugniet and
Faith Scheideman
Advisors and Proofreaders

Sandy
My Wife, Proponent and Ally

Fulda Gap

A Narrative Poems
written in
Rhyming Couplets

BY

DENNIS M KEATING

Dennis M Keating, the author of **The Fulda Gap,** has enjoyed a rather peripatetic life. His stories reflect this as each takes place in a different locale – Germany, Thailand, Hawaii, Texas and Chicago.

All five stories are true. Four relate to Keating's personal experiences. The fifth took place almost ninety years ago, but its initial incident occurred just a half block from Keating's current home.

The stories are written for male audiences as they include action, adventure and/or murder in their central themes. They are written in a poetic, rhyming couplet format. Hopefully, this will encourage more men to develop an interest in verse and thereby expand the realm of poetry.

While these tales include gritty elements, many women will also appreciate them. Trustfully, all audiences will find them interesting and compelling.

The
Fulda Gap

The Fulda Gap tells of the Cold War situation that existed in Germany from the late 1940's until 1990. *Fulda* is a small town in central Germany. The *Fulda Gap* is an open plains area that played a critical role during the tense Cold War. The Cold War ended with the destruction of the Berlin Wall, along with the 860 miles of landmines, barriers and guard posts that divided Germany from the Baltic Sea to the Czechoslovakia border.

In prior centuries invading army knew the Fulda Gap as the Hessian Corridor. During the Cold War, US Army soldiers of the 3rd Armored Division and other units stood eyeball to gun turret against the Soviet Block tanks. Our guys and gals where the first line of defense against the communist forces that sought to demolish Germany. During this perilous fifty years, the Third Herd took on the moniker, *Defenders of the Fulda Gap.*

Cold War: Inner German Border

West Germany

Helmstedt

Berlin

Fulda

Russian Tanks

East Germany

Frankfurt

Most people don't know
the Fulda Gap,

And couldn't find it
on a map.

It's located across
the Atlantic sea,

Smack dab in the middle
of Germany.

This open flatland,
half a world away,

Has been well traversed
since ancient day.

An American Sector sign, located near Berlin's Checkpoint Bravo, in English, Russian, French and German. To us, it meant, once you go passed this point, we cannot promise we can help you.

Its military significance
goes back to yore.

The Kaisers knew it
as The Hessian Corridor.

It's a wide plain that cuts

East to West.

Large armies can cross it

with ease and finesse.

Napoleon viewed the Gap

as a European main street

When his troops attacked Moscow,
then beat a fast retreat.

In 1812, Napoleon sought to place all of Europe under the control of France. With this goal, he sent the French Grande Armée to conquer Moscow. The campaign was a total failure that left Napoleon's military permanently crippled.

The Gap's character changed
after World War II.

You see, the Cold War
put Germany to the screw.

The old Vaterland
literally, was cut in half,

Like some evil demons
sliced it with a magic staff.

With this, central German towns

took on a border role,

And an East Bloc Commie monster
eyed them with evil in its soul.

Cold War: The half century of military and political hostility between the West and the Soviet Bloc.

Fulda: A small town in central Germany.

Fulda Gap: A central Germany lowlands area that is very suitable for the swift movement of military troops and tanks.

Hesse: A federal state in the Federal Republic of Germany.

Vaterland: German for Fatherland.

East Bloc: The group of communist states located in Central and Eastern Europe that were virtually under the control of Russia from after World War II until December 26, 1991, the day the USSR (Union of Soviet Socialist Republics) was officially dissolved.

Commie: A Communist, mainly a Soviet.

The town of Fulda
became the crosshair place,

Where soldiers on both sides
stared, face to face.

If Russian tanks were to attack,
there's no doubt

The lowland openings near Fulda
would be the route.

West Germans knew this probability

to be very great.

They'd studied Cold War history well.
And could only to sit and wait.

Soviets: Member of the former Russian controlled USSR (Union of Soviet Socialist Republics).

In the Fall of 1956, Soviet military and tanks invaded Hungary and its capital, Budapest.

Ivan: Common Russian name that became a nickname for a Communist or a Soviet.

Ruskie: Nickname for Russian.

In the Summer of 1968, Soviet military and tanks invaded Czechoslovakia and its capital Prague.

Klick: Slang for kilometer. Ten kilometers or klicks equals 6.2 miles.

Willkommen: Welcome in German.

ICBM: Intercontinental Ballistic Missile.

SAC: US Air Force Strategic Air Command.

In the 1956 attack on Budapest,

2,500 of Ivan's tanks led;

Three days later,

20,000 Hungarians lay dead.

Then, 2,000 Ruskie tanks
moved on Prague in '68,

The Czechs quickly surrendered
to avoid the same fate.

What options could
the Allied generals weigh?

All of West Germany was

Just a few hundred klicks away!

Cold War: Beginning and End

The war in Europe ended with the surrender of Germany on May 7, 1945. During the next two years, the Soviets, under Stalin, continued a policy of geopolitical aggression. On March 12, 1947, President Truman stated enough is enough. He announced the US would begin to counter the Soviets. This day is viewed as the day the Cold War started.

On November 9, 1989, the Berlin Wall was breached. This was not the end of the Cold War. While the opening of the wall was a very major incident, it was just one event in the whole domino process. Officially, the Cold War ended on December 26, 1991, when the USSR or Soviet Union totally broke apart, dissolved, and ceased to exist.

Please see Cold War Timelines at end of this poem.

Should a massive tank charge
plow through,

West Germany'd be flattened
in a day or two.

With armies fighting,
not only soldiers would die;

The civilian body count
would also be quite high.

When would the attack occur?
Well, Moscow held that ace.

We just knew, when it came,
Fulda would be the place.

A US M60 Tank

The NATO to plan,
you may not want to hear

For the Allied bosses moved
the German troops to the rear.

At each Fulda fence post
stood an American GI.

Yeah! Right up front! Our guys
were to be the first to die.

"Willkommen in Deutschland!
Enjoy your stay.

Hope a tank won't roll over you

before next Tuesday!"

The insignia of the US Army 3rd Armored Division, sometimes referred to as the Third Herd. During the Cold War, they were **The Defenders of the Fulda Gap**.

As an American, you might think,

"What the hey?"

Take a breather, bud.
This was just Part A.

If the West Germans were up front,
the Russians could shout,

"This is an internal fight.
Everyone else butt out."

But with Yanks taking the lead,
the US would say,

"If you kill one of our guys.
Big time, you're gonna pay."

Cold War: The City of Berlin

There is a misconception that Berlin was a border city during the Cold War. This is incorrect. Berlin was located some 110 inside Communist held East Germany. Berlin was totally encircled by East Germany. West Germany had two boundaries: the 850-mile barrier that separated West and East Germany; and the 96-mile wall that encircled Berlin.

At the end of World War II, Berlin was divided into four sectors overseen, respectively, by the British, French, Americans and Soviets. The three western powers cooperated and their sectors melded together. The Soviet sector linked with East Germany. On June 24, 1948, the Soviets blockaded all shipments into West Berlin. The Soviet goal: Starve to death all the West Berliners. The Allies countered this plan with the Berlin Airlift.

Part B of our plan came from Alaska,
Our 49th state,

It concerned our heavy hitters,
who could seal Russia's fate:

An ICBMs and long-range bombers mix,
courtesy of SAC.

They packed nuclear warheads.
A real and frightening fact.

Our Alaskan bomber wing planes
were always in the air.

Of this, the Kremlin leaders
were keenly aware.

Map of Cold War Berlin

East Germany

Berlin

French

British

Soviet

X
Checkpoint
Charlie

USA

X
Checkpoint
Bravo

Autobahn
to West
Germany

If USSR tanks rumbled thru Fulda,
on any given day,

Our missiles were already locked on
Moscow just four hours away.

This chilly standoff played out
for forty long years.

Both sides were tense, but they hid
their private fears.

Whenever I ventured near the fence,
I held a certain dread.

If a young Russian soldier screwed up.
then Blam! Blam! Everyone's dead.

Cold War: Orders for Travel to Berlin

We military-related Americans needed *Movement Orders* documents when driving through East Germany.

Note the *Movement Orders* and their dates on the next page. The *Orders* are in Russian, English and French, and have Russians approval stampings. Yes, the Russians had total control over East Germany; and did so, for a couple years after the Berlin Wall went down.

When we traveled by car, the American Military Police always cautioned us: Don't drive too fast, nor too slow. Drive too fast and you give the Soviets a reason to arrest you. Drive too slow and you fail to arrive at the predetermined time. Then, the US Command presumes the worst and sends a helicopter to pluck you out of Ivan's hair and return you to Uncle's care.

Sometimes, with official orders,
I'd cross through the fences.

Each time I did,
it numbed my senses.

The East and West differences
were beyond rad.

When I returned home,
my heart was empty and sad.

Just a few steps thru the checkpoint
I'd start to mull:

Why is East Germany so dreary,
desolate, and dull?

Movement Order/Travel Document

UNITED STATES OF AMERICA
ÉTATS-UNIS D'AMÉRIQUE
СОЕДИНЕННЫЕ ШТАТЫ АМЕРИКИ

MOVEMENT ORDERS
LAISSEZ-PASSER
ПУТЕВКА

Name Nom, Prénom Фамилия, Имя	Rank Qualité Чин	Nationality Nationalité Гражданство	Identity Document No. Pièce d'identité No. № удостоверения личности
KEATING DENNIS M	Civ	American	NU 65,528A
JOHN THOMPSON	Civ	American	NY 66,653A
BOGHOS MICHAEL L	Civ	American	NY 66,949A

is / are authorized to travel from
est / sont autorisé(s) à se rendre de
уполномочен/уполномочены
следовать из   Helmstedt   to / à   Berlin   and return / et retour / и обратно

by train or by vehicle No.
par le train ou par voiture No.
поездом или на автомашине №   KD 7015

from (date) / du (date) / от (числа)   17 Feb 90

to (date) / au (date) / по (число)   22 Feb 90   inclusive / inclus / включительно

The Commander-in-Chief of the United States Army, Europe
Le Commandant-en-Chef de l'Armée Américaine en Europe
Главнокомандующим Американской Армии в Европе.

Signature / Подпись

Title / Qualité / Звание   Lieutenant Colonel, Adjutant General

Date / Число   16 Feb 90

Document needed to travel through East Germany.

It's the same people, same families,
just a barrier fence apart.

But West Germany was prosperous;
the East was not.

Fortunately, this crazy powder keg
never did blow.

I feared the fuse might speed up!
But, it burned forever slow.

There were no tank attacks;
No fiery climax, nor a big boom.

Time just had its way of undercutting
the pending, certain doom.

Cold War: Travel to Berlin

When we drove to Berlin, our journey began at the Helmstedt border crossing, **Checkpoint Alpha**. From there, we took Bundesautobahn 2, for 110 miles, or 180 Klicks, until we got to **Checkpoint Bravo** the American entrance to West Berlin.

Driving through East Germany could be a bit unnerving, like when we encountered a soviet tank or troop carrier. For sure, we had to stop at a few foreboding East German checkpoints. Sometimes, the checkpoint guards hassled us. At other times, they wanted to swap black market goods for money. Were these legit offers? Or, were we being set up for a sting? To me, indefinite lodging in a Soviet cell didn't sound cool, so I forewent the offers and never found out. In hindsight, I preferred flying from Munich to Berlin's Tempelhof airport. It was more relaxing.

It took four decades for the Commie

Façade to wear thin;

Corruption and stagnation ate
the Soviet Union from within.

Finally, in 1989, the Berlin Wall

came tumbling down,

Freedom, joy, and peace became
the new game in town.

Those involved felt great relief

from the Cold War strain;

And very few have an interest

In doing a replay, again.

Cold War: Checkpoint Charlie

The famous **Checkpoint Charlie** was used when going from West Berlin into Soviet Controlled East Berlin. When driving alone, **Charlie** could be a very harrowing Checkpoint. First, upon entrance, an East German soldier would thrust a Kalashnikov at my face. My orders: ***Don't roll down the window. Just show your Travel Orders through the glass.*** After a few moments, the guard would back down and lower his weapon.

Then, after maneuvering my auto through several concrete barriers, two Russian tanks greeted me. The tanks would adjust their gun turrets so that they aimed directly at my front windshield. Then, after a few moments I was waved though.

Believe me, I always felt great relief, upon my return to West Berlin, when I saw our Stars and Stripes flying in the breeze.

While the Fulda Gap fence isn't
as well known as the Berlin Wall,

To those trapped inside,
that fact meant nothing at all.

For them, both barriers
were visible to see,

And both existed to keep
those inside, other than free.

Sure, this crazy period ended,
More than twenty-five years ago

But due to its impact on our world
It's history that's good to know.

Old Glory

When one travels throughout the world and encounters hotspots and challenging situations, it is always refreshing to come upon our Red, White and Blue flying high over one of our embassies, consulates or military installations.

Background: Pre-World War II

For several centuries, Europe consisted of kingdoms, big and small. Due to changing alliances, invasions and military actions, the various boundaries frequently shifted and changed.

World War I ended with the Treaty of Versailles. This resolved many matters, but also created new ones. Naturally, the winners wrote the treaty for their advantage, while the losers, mainly, the Germans, were left with bitterness. A decade later, Hitler would build a power base that drew upon the festering wounds felt by his countrymen.

Meanwhile, Poland and other countries were caught between the super powers of Germany and Russia. Poland became a punching bag with its shape being continually transformed by its neighbors.

The author at the Berlin Wall. With sledge hammer in hand, he had the distinct personal pleasure of participating in the destruction of the Berlin Wall. He still keeps his souvenirs.

Cold War Background: Adolph Hitler

1889 Apr 20 Hitler born in Austria, across the river from Germany.

1914 Jul 28 World War I starts; Hitler joins the Bavarian Army.

1919 Sep 12 Hitler joins DAP Workers' Party and becomes popular orator. Within six months, the DAP becomes the Nazi party. Within two years, Hitler becomes leader of Nazi party with his Make Germany Great Again rallies.

1923 Nov 11 Hitler arrested for high treason after failed Beer Hall Putsch where he attempted to overthrow the government. While in prison, Hitler writes *Mein Kampf*.

1933 Jan 30 Hitler is German Chancellor.

1934 Aug 02 Hitler becomes **Der Führer**.

Cold War Background: Joseph Stalin

1912 Jan 01 Vladimir Lenin founds the Russian Communist Party.

1917 - 1918 Lenin's communist forces overthrow the Czarist government of Russia. Within a few months, Lenin has the Czar, his wife and children executed.

1922 Apr 03 Stalin rises to Party General Secretary developing strong power base.

Dec 30 The USSR (Union of Soviet Socialist Republics) is formed. It grows to 16 republics with Russia at the helm. A sickly Lenin becomes the Top Banana, but Stalin starts taking major power grabs.

1923 Mar 09 Lenin has massive stroke.

1924 January 21. Lenin dies. Communist Party infighting intensifies with the two main players being Stalin and Trotsky.

1928 Jan 31 Stalin exiles Trotsky. Within a year, Stalin takes total control and becomes absolute dictator. Trotsky, goes on the run, relocating from country to country with Stalin Hit Squad keeping tab. In August 1940, Trotsky is assassinated.

1929 Dec 29 Stalin speech attacks the Kulaks, the wealthier farmers, signaling the start of Stalin' brutal Reign of Terror. Over the next two decades, many millions will be executed or forced into forced labor camp Gulags.

1932 Nov 09 Stalin's wife's suicide.

1941 Jun 22 Hitler launches **Operation Barbarossa**, bombing Soviet controlled Poland and sending troops to attack Russia. Stalin is forced to refocus and fight Hitler's troops. The German-Soviet part of World War II begins.

1953 Mar 03 Stalin dies.

Cold War: Soviet Power

There are recurring questions concerning how the Soviets obtained so much control and unbridled power over East Germany and Eastern Europe during the forty years of the Cold War. Much of this answer can be tied to the Tehran Conference in 1943 and the Yalta Conference of 1945. Many historians believe, at these conferences, the Soviet leader, Joseph Stalin, simply bamboozled, snookered and outfoxed the American leader, Roosevelt, who was too vain, ignorant and naïve to understand the long-range goals of the Soviets. While historians may disagree, the reality is, these two conferences played a major role in sentencing to death and/or long term slavery like conditions, innumerous peoples in both Eastern Europe and East Germany. A Cold War chronology follows.

Cold War Background: World War II

1936 Oct 25 Germany & Italy form Axis. Nov 25 Germany Japan Pact to fight USSR.

1939 Sep 01 Germany, under Hitler, invades Poland starting World War II.

Sep 03 Britain, France and others declare War on Germany.

Sep 17 Russia invades Poland.

1940 Sep 27 Japan joins Axis.

1941 Jun 22 Germany attacks Russia. Dec 06 Germany is pushed back from attack on Moscow (USSR).

Dec 07 Japan attacks Pearl Harbor. USA enters war, greatly expand World War II.

Dec 11 Germany declare war on USA.

1942 Apr 18 Doolittle Raid. US B-25 bombers attack Tokyo. The raid is a morale booster and sends a "Remember Pearl Harbor" to Japanese leaders.

1943 Nov 28 **Tehran Conference** (Churchill: Great Britain; Roosevelt: USA; Stalin: USSR) Held at Soviet Embassy. Roosevelt stays in a Stalin provided room, that is probability (99.9%) bugged.

1944 Jun 06 **D-Day** USA, Britain and Canada invade the European mainland at Normandy, France to repel the Germans. Jul 25 Allied Forces breakthrough Normandy and start to move forward. Aug 20 Allied troops enter Paris.

Oct 27 The 442nd Infantry Regiment is ordered to rescue the trapped Texas *Lost Battalion* near the German border. The all Japanese-American 442nd, starts with 4,000 men. It outfights strategically and numerically superior enemy forces with its Hawaii pidgin *Go For Broke* spirit and goes on to become the most decorated unit in the history of American warfare. The 442 earned 9,486 Purple Hearts and 21 Medals of Honors.

Dec 16 Germany's last major offensive, the *Battle of the Bulge* begins.

World War II Ends - Cold War Begins

1945 Jan 15 Allies advance on Berlin. Hitler relocates his headquarters to the Führerbunker, an underground bunker. Feb 04 **Yalta Conference** - Codename: *Argonaut.* (Churchill, Roosevelt and Stalin). Again, Stalin controls the show. Feb 13 Allies firebomb Dresden. Mar 09 US firebombs Tokyo. Probably, the deadliest bomb attack in history. April 12 Roosevelt dies. Truman becomes US president.

April 20 Hitler celebrates 56th birthday as Soviet forces encircle Berlin. April 29 Hitler marries his longtime girlfriend, Eva Braun. Americans liberate Dachau Concentration Camp near Munich. April 30 Hitler and his new wife start their honeymoon by committing suicide. May 01 A German general negotiates a surrender of Berlin to Soviets. May 02 Soviets capture the Reichstag, Germany's capitol building and raise Hammer & Sickle flag atop the roof.

May 07 Germany surrenders to Allies. V-E Day (Victory in Europe Day)

June 05 Germany is divided into quarters, (American, British, French and Soviet.) July 17 **Potsdam Conference** (Churchill, Stalin and Truman). Potsdam, a suburb of Berlin. Goal: to settle World War II matters. Churchill is replaced by Clement Attlee due to British election. Truman, unlike the fawning Roosevelt and with A-bomb plan up his sleeve, matches Stalin's poker face resolve. Stalin cannot bully Truman, nothing is settled.

Jul 26 With the **Potsdam Proclamation** Truman tells Japan: Make an immediate unconditional surrender or face "prompt and utter destruction." America drops pamphlets on Japanese cities warning of more bombs to come.

Jul 28 Japan ignores Truman with a screw you, Mokusatsu (kill it with silence) reply. Aug 02 Potsdam Conference Ends. Aug 06 Americans drop Little Boy on Hiroshima killing 70,000 Japanese.

Aug 08: Japanese leaders don't surrender. Stalin declares war on Japan.

Aug 09: Americans drop Fat Man and kill 80,000 in Nagasaki, Japan.

Aug 09 After six years of neutrality, Soviet troops invade Japanese held Manchuria.

Aug 10 Japan gives in and accepts Potsdam unconditional surrender terms.

Aug 15: V-J Day (Victory over Japan Day.)

Aug 24 Soviet Soldiers march into Korea. Sep 02 Japan officially surrenders aboard battleship USS Missouri in Tokyo Harbor. Soviets end 24-day war with Japan.

Dec 31 12:00 noon. **Proclamation 2714**. World War II officially ends.

1946 Feb 08 Kim Il-Sung becomes boss of Soviet Controlled (northern) Korea.

Feb 09 Stalin declares war is inevitable due to the dissimilarities of communism and capitalism.

Mar 05 Churchill's "Iron Curtain" speech.

The Cold War

The Cold War was a West (USA, NATO and its allies) verse East (Russia, the USSR and other communist states) standoff that began at the end of World War II and did not end until the USSR dissolved in 1991. While it impacted the world, Germany often proved to be the playing field.

1947 Mar 12 Truman announced US is prepared to counter the ongoing Soviet geopolitical aggression.

Jun 05 US announces Marshal Plan

1948 Feb 21 Czechoslovakia. Overthrown by Soviet Supported Communist forces. Jun 18 US, Britain and France agree on a common currency, the Deutsche Mark, for Germany. USSR says "No." This in effect establishes West Germany.

Jun 24 Soviets blockade Berlin.

Jun 25 Berlin Airlift begins.

1949 Apr 04 Creation of NATO.

May 12 USSR ends blockade of Berlin. The Berlin airlift continued in order to build up a sufficient stockpile of goods so that the city could function as normal.

Aug 29 Soviets explodes Nuclear Bomb. Sep 30 Berlin Airlift ends. Last plane carries two tons of coal for the winter. Oct 01 People's Republic of China established under Mao Zedong. Oct 24 United Nations established.

1950 Jun 25 Korean War begins with North Korea invading the South. Sep 15 US forces land at Inchon to counter North Korea.

1953 Jul 27 Korean Armistice signed.

1960 May 01 USSR missile shoots down Gary Powers' U-2 spy plane. Powers is taken prisoner.

1961 Apr 17 Bay of Pigs Invasion in Cuba. Aug 13 Just after midnight, East Germany seals border with West Berlin and begins construction of the Berlin Wall preventing East Germans to entry West Berlin.

1962 Feb 10 U-2 Pilot Powers released and swapped for a Russian spy at Berlin's Bridge of Spies (Glienicke Bridge).

Oct 16 Cuban Missile Crisis US halts Russian missiles plan.

1963 Jun 23 US President Kennedy visits West Berlin and, gives his famous *Ich Bin Ein Berliner* speech.

1964 Aug 02 Vietnam Gulf of Tonkin Incident, causing large scale US troop involvement in the Vietnam war.

1975 Apr 30 Saigon falls and the US exits Vietnam.

1985 Mar 11 Mikhail Gorbachev becomes top dog in the USSR. Between Stalin and Gorbachev there were five other leaders. Gorbachev realizes he has inherited a bloated, corrupt and highly inefficient bureaucracy. To counter this reality, he introduces Perestroika (restructuring) and Glasnost (openness) and decides to end the economic aid to Soviet satellites and open a way for them to break away from the USSR.

1986 Apr 26 Chernobyl nuclear disaster. Dec 26. Riga, Latvia. Early morning, after attending a rock concert several hundred youths march downtown shouting *Soviet Russia out! Free Latvia!* This incident causes waves of unrest throughout the USSR as youths in the other soviet satellites were inspired to demonstrate against Russia.

1987 Jun 12 President Reagan, in front of the Brandenburg Gate and the Berlin Wall gives his famous, *Mr. Gorbachev, tear down this wall* speech.

1989 Aug 19 Hungary opens border to Austria and allows 13,000 East German tourists to escape to the West. Nov 09 Berlin Wall breached and East Germans can freely move into the West.

1990 May 04 Latvia gets independence from USSR. Other republics follow suit.

Oct 03 The two Germany's are reunited. Oct 15 Gorbachev gets Nobel Peace Prize.

1991 Dec 26 Soviet Union dissolved and the Cold War basically ended. Game over. The Good Guys Won. Everyone go home. The various American military units and the Department of Defense activities began a serious review and phase down of all their European Operations.

1992 Jan 21 Four weeks later, the author, at age 51, was offered early retirement. offer. He knew his job was done. There was nothing more to do. It was time to move on. With a simple "Thank you and Auf Wiedersehen," he retired to a new life.

Thanks to his unique job circumstances during the last few years of the Cold War, the author was one of the few non-military authorized entry to every American military encampment along the East West German border. Prior to this, he had also been allowed entry to the top-secret SAC Control Center at McChord AFB in Washington state. The Control Center monitored our nuclear bombers as they headed toward Moscow. When the bombers reached the North Pole, the Control Center would order them to return to their base in Alaska. Immediately, another group would be instructed to replace them and directed to head to the North Pole.

The German government was quite appreciative of the author's service in Europe. Upon the author's retirement, Germany offered the author full German citizenship; permanent residence; and a four-bedroom home in Munich at a very low rental rate. He kept his Munich apartment until 2005, thirteen years after his retirement. At that time, he chose to come out of the cold and return home to the warmth of Honolulu. The author continues to have a strong love for Germany and the German people.

East meets West at Berlin Wall Opening. The author, Dennis M Keating, after doing a little damage to the Berlin Wall, shakes hands with an East German soldier.

Thank You
and
Auf
Wiedersehen

ABOUT THE AUTHOR

Dennis M Keating has enjoyed a peripatetic lifestyle. His international perspective and eclectic enthusiasm for life come from his forty some years in Germany; Thailand; China and Hawaii.

For the last ten years, Keating and his wife, Sandy, have been living a quiet life in Waikiki. Normally, he can be found pounding his iMac keyboard, hiking the Diamond Head trail, or strolling with his wife at sunset along the sands of Waikiki.

Keating writes on a diverse range of topics. His books draw upon his multifarious interests and personal experiences. Most of his books are nonfiction.

Keating's Facebook page:
https://www.facebook.com/TheHonoluluGuy/
He is happy to Friend you on Facebook.

In 2016, Keating released - *The Olympics: An Unauthorized Unsanctioned History*

In 2017, Keating released
Poetry for Men - Action Adventure Murder is a compilation of Keating's five poetry books.

Charlie Whitman was a Friend of Mine. The story of the Texas Tower Killer.

Ena Road. Murder and racism in Hawaii.

The Fulda Gap. A Cold War confrontation.

A Chicago Tale. A triple murder story.

Black Lahu. Opium, life and death in the Golden Triangle.

His email is **lostpuka@gmail.com**
His websites are:
GoldenSphere.com & **HonoluluGuy.com**

Keating owns all rights to the material in this book. For film rights, or for other reasons, please contact him.

Ciao!